STONER Coloring Book

Color When HIGH

How HIGH Am I?

Time: _____   Date: _____

Location: _____

- ☐ Not feelin' it
- ☐ Here it comes
- ☐ Buzzing
- ☐ This is it
- ☐ Where's the food?
- ☐ I'm cooked
- ☐ What did you say?
- ☐ What was I saying?
- ☐ Couch potato
- ☐ I'm definetely too high

How am I feeling? _____
_____
_____

What am I doing/eating? _____
_____
_____

High thoughts _____
_____
_____

How HIGH Am I?

Time: _____   Date: _____

Location: _____

- ☐ Not feelin' it
- ☐ Here it comes
- ☐ Buzzing
- ☐ This is it
- ☐ Where's the food?
- ☐ I'm cooked
- ☐ What did you say?
- ☐ What was I saying?
- ☐ Couch potato
- ☐ I'm definetely too high

How am I feeling? _____
_____
_____

What am I doing/eating? _____
_____
_____

High thoughts _____
_____
_____

How HIGH Am I?

Time: _____ Date: _____

Location: _____

- [ ] Not feelin' it
- [ ] Here it comes
- [ ] Buzzing
- [ ] This is it
- [ ] Where's the food?
- [ ] I'm cooked
- [ ] What did you say?
- [ ] What was I saying?
- [ ] Couch potato
- [ ] I'm definetely too high

How am I feeling? _____

_____

_____

What am I doing/eating? _____

_____

_____

High thoughts _____

_____

_____

How HIGH Am I?

Time: _____   Date: _____

Location: _____

- ☐ Not feelin' it
- ☐ Here it comes
- ☐ Buzzing
- ☐ This is it
- ☐ Where's the food?
- ☐ I'm cooked
- ☐ What did you say?
- ☐ What was I saying?
- ☐ Couch potato
- ☐ I'm definetely too high

How am I feeling? _____
_____
_____

What am I doing/eating? _____
_____
_____

High thoughts _____
_____
_____

How HIGH Am I?

Time: _____ Date: _____

Location: _____

- [ ] Not feelin' it
- [ ] Here it comes
- [ ] Buzzing
- [ ] This is it
- [ ] Where's the food?
- [ ] I'm cooked
- [ ] What did you say?
- [ ] What was I saying?
- [ ] Couch potato
- [ ] I'm definetely too high

How am I feeling? _____
_____
_____

What am I doing/eating? _____
_____
_____

High thoughts _____
_____
_____

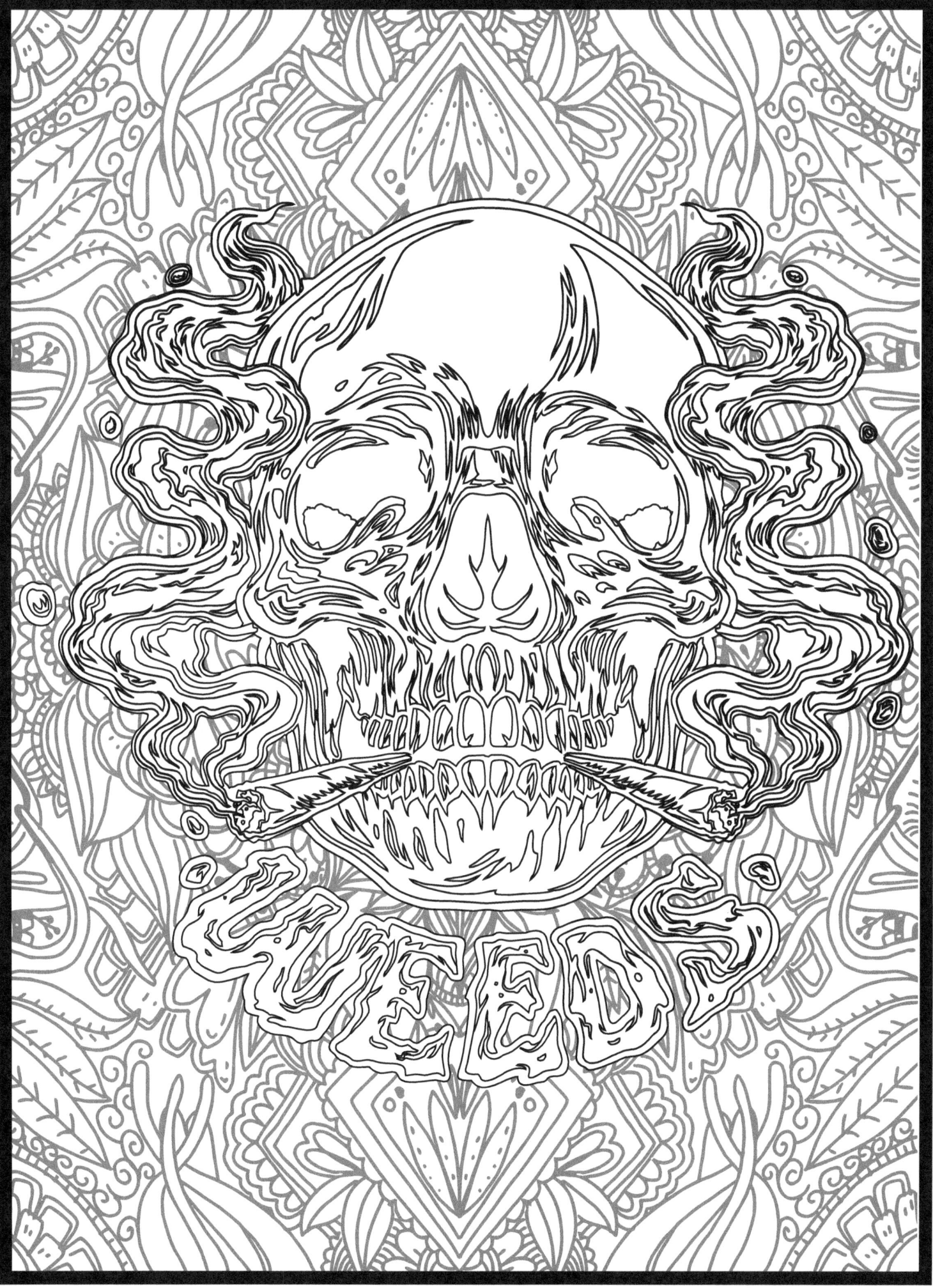

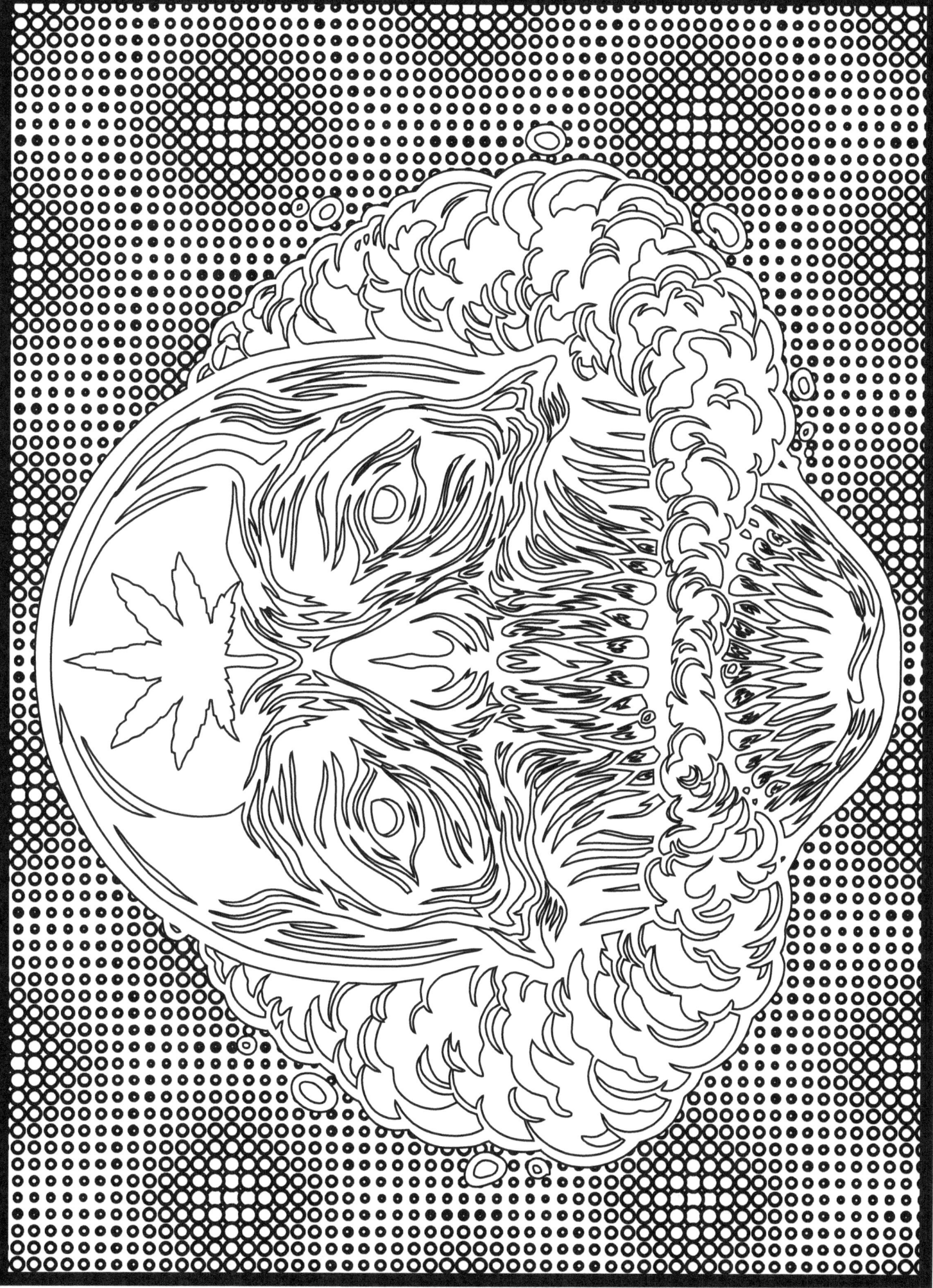

What Did You Think of That Coloring Book?

First of all, thank you for purchasing this book. We know you could have picked any number of books, but you picked this one and for that we are extremely grateful.

If you enjoyed this book and found some benefit in it, we'd like to hear from you and hope that you could take some time to post a review on Amazon. Your feedback and support will help us to greatly improve our designing for future projects and make this book even better.

We wish you all the best !!!

www.ingramcontent.com/pod-product-compliance
Lightning Source LLC
Chambersburg PA
CBHW082018230526
45466CB00022B/2574